ICE Conditions of Contract for Minor Works

Second Edition

CONDITIONS OF CONTRACT, AGREEMENT AND CONTRACT SCHEDULE FOR USE IN CONNECTION WITH MINOR WORKS OF CIVIL ENGINEERING CONSTRUCTION

Institution of Civil Engineers
Association of Consulting Engineers
Federation of Civil Engineering Contractors

Published for The Institution of Civil Engineers, the Association of Consulting Engineers and the Federation of Civil Engineering Contractors by Thomas Telford Services Ltd, Thomas Telford House, 1 Heron Quay, London E14 1JD

Copies may also be obtained from
The Director General and Secretary
The Institution of Civil Engineers
1–7 Great George Street
London SW1P 3AA

The Secretary
The Association of Consulting Engineers
Alliance House
12 Caxton Street
London SW1H 0QL

The General Secretary
The Federation of Civil Engineering Contractors
Cowdray House
6 Portugal Street
London WC2A 2HH

First edition 1988
Second edition 1995

The Institution of Civil Engineers, the Association of Consulting Engineers and the Federation of Civil Engineering Contractors have, as sponsoring authorities, approved this revised second edition of the document commonly known as the ICE Conditions of Contract for Minor Works, for all minor works of civil engineering construction. A permanent joint committee of the sponsoring authorities will keep under review the use of this document and will consider any suggestions for amendment, which should be addressed to the Director General and Secretary, the Institution of Civil Engineers, 1–7 Great George Street, London SW1P 3AA. Revision to the document will be made when such action seems warranted.

A CIP cataloguing record for this book is available from the British Library

ISBN 0 7277 2037 6

Printed and bound in Great Britain by Spottiswoode Ballantyne Printers Ltd., Colchester

CONTENTS

INDEX

ICE Conditions of Contract for Minor Works

AGREEMENT

THIS AGREEMENT is made the .... day of .... 19 ....

between ....

of (or whose registered office is at) ....

....

(hereinafter called the 'Employer') of the one part

and ....

of (or whose registered office is at) ....

....

(hereinafter called the 'Contractor') of the other part

WHEREAS the Employer wishes to have carried out the following ....

....

....

and has accepted a Tender by the Contractor for the same

NOW IT IS HEREBY AGREED AS FOLLOWS:

Article 1 The Contractor will subject to the Conditions of Contract perform and complete the Works.

Article 2 The Employer will pay the Contractor such sum or sums as shall become payable under the Contract and in accordance with the Conditions of Contract.

Article 3 The documents listed in the Contract Schedule form part of this Agreement.

AS WITNESS the hands of the parties hereto:

Signed for and on behalf of the Employer ....

in the presence of ....
(Witness)

Signed for and on behalf of the Contractor ....

in the presence of ....
(Witness)

1

ICE Conditions of Contract for Minor Works

CONTRACT SCHEDULE
(List of documents forming part of the Contract)

The Agreement (if any)

The Contractor's Tender (excluding any general or printed terms contained or referred to therein unless expressly agreed in writing to be incorporated in the Contract)

The Conditions of Contract

The Appendix to the Conditions of Contract

The Drawings. Reference numbers ....

....

....

The Specification. Reference ....

The priced Bill of Quantities*

The Schedule of Rates*

The Daywork Schedules*

The following letters*

from .... to .... dated ....

from .... to .... dated ....

from .... to .... dated ....

from .... to .... dated ....

*Delete if not applicable.

2

ICE Conditions of Contract for Minor Works

1 DEFINITIONS

Definitions **1.1** "Works" means all the work necessary for the completion of the Contract including any variations ordered by the Engineer.

1.2 "Contract" means the Agreement if any together with these Conditions of Contract the Appendix and other items listed in the Contract Schedule.

1.3 "Cost" (except for "cost plus fee" contracts, see Appendix) means all expenditure properly incurred or to be incurred whether on or off the Site including overhead finance and other charges (including loss of interest) properly allocatable thereto but does not include any allowance for profit.

1.4 "Site" means the lands and other places on under in or through which the Works are to be constructed and any other lands or places provided by the Employer for the purposes of the Contract together with such other places as may be designated in the Contract or subsequently agreed by the Engineer as forming part of the Site.

1.5 "Excepted Risks" are

(a) the use or occupation by the Employer his agents servants or other contractors (not being employed by the Contractor) of any part of the Permanent Works

(b) any fault defect error or omission in the design of the Works (other than a design provided by the Contractor pursuant to his obligations under the Contract)

(c) riot war invasion act of foreign enemies or hostilities (whether war be declared or not)

(d) civil war rebellion revolution insurrection or military or usurped power

(e) ionizing radiations or contamination by radioactivity from any nuclear fuel or from any nuclear waste from the combustion of nuclear fuel radioactive toxic explosive or other hazardous properties of any explosive nuclear assembly or nuclear component hereof

and

(f) pressure waves caused by aircraft or other aerial devices travelling at sonic or supersonic speeds.

2 ENGINEER

Engineer to be a named individual **2.1** The Employer shall appoint and notify to the Contractor in writing a named individual to act as the Engineer. If at any time the Engineer is unable to continue the duties required by the Contract the Employer shall forthwith appoint a replacement and shall so notify the Contractor in writing.

2.2   The Engineer may appoint a named Resident Engineer and/or other suitably experienced person to watch and inspect the Works and the Engineer may delegate to such person in writing any of the powers of the Engineer herein provided that prior notice in writing is given to the Contractor.

Engineer's power to give instructions   **2.3**   The Engineer shall have power to give instructions for

(a) any variation to the Works including any addition thereto or omission therefrom

(b) carrying out any test or investigation

(c) the suspension of the Works or any part of the Works in accordance with Clause 2.6

(d) any change in the intended sequence of the Works

(e) measures necessary to overcome or deal with any obstruction or condition falling within Clause 3.8

(f) the removal and/or re-execution of any work or materials not in accordance with the Contract

(g) the elucidation or explanation of any matter to enable the Contractor to meet his obligations under the Contract

(h) the exclusion from the Site of any person employed thereon which power shall not be exercised unreasonably.

2.4   The Engineer or Resident Engineer and/or other suitably experienced person who exercises any delegated power shall upon the written request of the Contractor specify in writing under which of the foregoing powers any instruction is given. If the Contractor is dissatisfied with any such instruction he shall be entitled to refer the matter to the Engineer for his decision.

Dayworks   **2.5**   The Engineer may order in writing that any work shall be executed on a daywork basis. Subject to the production of proper records the Contractor shall then be entitled to be paid in accordance with a Daywork Schedule included in the Contract or otherwise in accordance with the "Schedules of Dayworks carried out incidental to Contract Work" issued by the Federation of Civil Engineering Contractors and current at the date the work is carried out.

Engineer may suspend the progress of the Works   **2.6**   (1) The Engineer may order the suspension of the progress of the Works or any part thereof

(a) for the proper execution of the work

(b) for the safety of the Works or any part thereof

(c) by reason of weather conditions

and in such event may issue such instructions as may in his opinion be necessary to protect and secure the Works during the period of suspension.

(2) If permission to resume work is not given by the Engineer within a period of 60 days from the date of the written Order of Suspension then the Contractor may serve a written notice on the Engineer requiring permission to proceed with the Works within 14 days from the receipt of such notice. Subject to the Contractor not being in default under the Contract the Engineer shall grant such permission and if such permission is not granted the Contractor may by a further written notice served on the Engineer elect to treat the suspension where it affects a part of the Works as an omission under Clause 2.3(a) or where the whole of the Works is suspended as an abandonment of the Contract by the Employer.

Parties bound by Engineer's instructions 2.7 Each party shall be bound by and give effect to every instruction or decision of the Engineer unless and until either

(a) it is altered or amended by an agreed settlement following a reference under Clause 11.3 and neither party gives notice of dissatisfaction therewith

or

(b) it is altered or amended by a decision of an arbitrator under Clause 11.4 or 11.5.

3 GENERAL OBLIGATIONS

Contractor to perform and complete the Works 3.1 The Contractor shall perform and complete the Works and shall (subject to any provision in the Contract) provide all supervision labour materials plant transport and temporary works which may be necessary therefor.

Responsibility for care of the Works 3.2 (1) The Contractor shall take full responsibility for the care of the Works from the starting date until 14 days after the Engineer issues a Certificate of Practical Completion for the whole of the Works pursuant to Clause 4.5.

(2) If the Engineer issues a Certificate of Practical Completion in respect of any part of the Works before completion of the whole of the Works the Contractor shall cease to be responsible for the care of that part of the Works 14 days thereafter and the responsibility of its care shall then pass to the Employer.

(3) The Contractor shall take full responsibility for the care of any outstanding work which he has undertaken to finish during the Defects Correction Period until such outstanding work is complete.

Contractor to repair and make good 3.3 (1) In case any damage loss or injury from any cause whatsoever (save and except the Excepted Risks) shall happen to the Works or any part thereof while the Contractor is responsible for their care the Contractor shall at his own cost repair and make good the same so that at completion the Works shall be in good order and condition and conform in every respect with the requirements of the Contract and the Engineer's instructions.

(2) To the extent that any damage loss or injury arises from any of the Excepted Risks the Contractor shall if required by the Engineer repair and make good the same at the expense of the Employer.

(3) The Contractor shall also be liable for any damage to the Works occasioned by him in the course of any operations carried out by him for the purpose of completing outstanding work or complying with his obligations under Clauses 4.7 and 5.2.

Contractor's authorized representative 3.4 The Contractor shall notify the Engineer of the person duly authorized to receive instructions on behalf of the Contractor.

Setting out and safety of site operations 3.5 The Contractor shall take full responsibility for the setting out of the Works and for the adequacy stability and safety of his site operations and methods of construction.

5

Engineer to provide necessary information 3.6 Subject to Clause 3.5 the Engineer shall be responsible for the provision of any necessary instructions drawings or other information.

Contractor's responsibility for design 3.7 (1) The Contractor shall not be responsible for the design of the Works except where expressly stated in the Contract.

(2) The Contractor shall be responsible for the design of any temporary works other than temporary works designed by the Engineer.

(3) The Contractor shall exercise all reasonable skill care and diligence in designing any part of the Permanent Works for which he is responsible.

Adverse physical conditions and artificial obstructions—delay and extra cost 3.8 (1) If during the carrying out of the Works the Contractor encounters any artificial obstruction or physical condition (other than a weather condition or a condition due to weather) which obstruction or condition could not in his opinion reasonably have been foreseen by an experienced contractor the Contractor shall as early as practicable give written notice thereof to the Engineer.

(2) If in the opinion of the Engineer such obstruction or condition could not reasonably have been foreseen by an experienced contractor then the Engineer shall certify a fair and reasonable sum and the Employer shall pay such sum to cover the cost of performing any additional work or using any additional plant or equipment together with a reasonable percentage addition in respect of profit as a result of

   (a) complying with any instructions which the Engineer may issue

and/or

   (b) taking proper and reasonable measures to overcome or deal with the obstruction or condition in the absence of instructions from the Engineer

together with such sum as shall be agreed as the additional cost to the Contractor of the delay or disruption arising therefrom. Failing agreement of such sums the Engineer shall determine the fair and reasonable sum to be paid.

Facilities for other contractors 3.9 The Contractor shall in accordance with requirements of the Engineer afford reasonable facilities for any other contractor employed by the Employer and for any other properly authorized authority employed on the Site.

4 STARTING AND COMPLETION

Starting date to be notified in writing 4.1 The starting date shall be the date specified in the Appendix or if no date is specified a date to be notified by the Engineer in writing being within a reasonable time and in any event within 28 days after the date of acceptance of the Tender. The Contractor shall begin the Works at or as soon as reasonably possible after the starting date.

Period for completion 4.2 The period or periods for completion shall be as stated in the Appendix or such extended time as may be granted under Clause 4.4 and shall commence on the starting date.

Contractor's programme 4.3 The Contractor shall within 14 days after the starting date if so required provide a programme of his intended activities. The Contractor shall at all times proceed with the Works with due expedition and reasonably in accordance with his programme or any modification thereof which he may provide or which the Engineer may request.

Extension of period for completion **4.4** If the progress of the Works or any part thereof is delayed for any of the following reasons

(a) an instruction given under Clause 2.3(a)(c) or (d)

(b) an instruction given under Clause 2.3(b) where the test or investigation fails to disclose non-compliance with the Contract

(c) encountering an obstruction or condition falling within Clause 3.8 and/or an instruction given under Clause 2.3(e)

(d) delay in receipt by the Contractor of necessary instructions drawings or other information

(e) failure by the Employer to give adequate access to the Works or possession of land required to perform the Works

(f) delay in receipt by the Contractor of materials to be provided by the Employer under the Contract

(g) exceptional adverse weather

(h) other special circumstances of any kind whatsoever outside the control of the Contractor

then provided that the Contractor has taken all reasonable steps to avoid or minimize the delay the Engineer shall upon a written request by the Contractor promptly by notice in writing grant such extension of the period for completion of the whole or part of the Works as may in his opinion be reasonable. The extended period or periods for completion shall be subject to regular review provided that no such review shall result in a decrease in any extension of time already granted by the Engineer.

Certificate of Practical Completion of Works or part of Works **4.5** (1) Practical completion of the whole of the Works shall occur when the Works reach a state when notwithstanding any defect or outstanding items therein they are taken or are fit to be taken into use or possession by the Employer.

(2) Similarly practical completion of part of the Works may also occur but only if it is fit for such part to be taken into use or possession independently of the remainder.

(3) The Engineer shall upon the Contractor's request promptly certify in writing the date on which the Works or any part thereof has reached practical completion or otherwise advise the Contractor in writing of the work necessary to achieve such completion.

Liquidated damages **4.6** If by the end of the period or extended period or periods for completion the Works have not reached practical completion the Contractor shall be liable to the Employer in the sum stated in the Appendix as liquidated damages for every week (or *pro rata* for part of a week) during which the Works so remain uncompleted up to the limit stated in the Appendix. Similarly where part or parts of the Works so remain uncompleted the Contractor shall be liable to the Employer in the sum stated in the Appendix reduced in proportion to the value of those parts which have been certified as complete provided that the said limit shall not be reduced.

Provided that if after liquidated damages have become payable in respect of any part of the Works the Engineer issues a variation order under Clause 2.3(a) or an artificial obstruction or physical condition within the meaning of

Clause 3.8 is encountered or any other situation outside the Contractor's control arises any of which in the opinion of the Engineer results in further delay to that part of the Works

(a) the Engineer shall so inform the Contractor and the Employer in writing

and

(b) the Employer's further entitlement to liquidated damages in respect of that part of the Works shall be suspended until the Engineer notifies the Contractor and the Employer in writing that the further delay has come to an end.

Such suspension shall not invalidate any entitlement to liquidated damages which accrued before the period of delay started to run and any monies deducted or paid in accordance with this Clause may be retained by the Employer without incurring interest thereon under Clause 7.8.

Rectification of defects 4.7 The Contractor shall rectify any defects and complete any outstanding items in the Works or any part thereof which reach practical completion promptly thereafter or in such a manner and/or time as may be agreed or otherwise accepted by the Engineer. The Contractor shall maintain any parts which reach practical completion in the condition required by the Contract until practical completion of the whole of the Works fair wear and tear excepted.

5 DEFECTS

Definition of Defects Correction Period 5.1 "Defects Correction Period" means the period stated in the Appendix which period shall run from the date certified as practical completion of the whole of the Works or the last period thereof.

Cost of remedying defects 5.2 If any defects appear in the Works during the Defects Correction Period which are due to the use of materials or workmanship not in accordance with the Contract the Engineer shall give written notice thereof and the Contractor shall make good the same at his own cost.

Remedy for Contractor's failure to correct defects 5.3 If any such defects are not corrected within a reasonable time by the Contractor the Employer may after giving 14 days written notice to the Contractor employ others to correct the same and the cost thereof shall be payable by the Contractor to the Employer.

Engineer to certify completion 5.4 Upon the expiry of the Defects Correction Period and when any outstanding work notified to the Contractor under Clause 5.2 has been made good the Engineer shall upon the written request of the Contractor certify the date on which the Contractor completed his obligations under the Contract to the Engineer's satisfaction.

Unfulfilled obligations 5.5 Nothing in Clause 5 shall affect the rights of either party in respect of defects appearing after the Defects Correction Period.

6 ADDITIONAL PAYMENTS

Engineer to determine additional sums and deductions 6.1 If the Contractor carries out additional works or incurs additional cost including any cost arising from delay or disruption to the progress of the Works as a result of any of the matters referred to in paragraphs (a) (b) (d) (e) or (f) of Clause 4.4 the Engineer shall certify and the Employer shall pay to the

Contractor such additional sum as the Engineer after consultation with the Contractor considers fair and reasonable. Likewise the Engineer shall determine a fair and reasonable deduction to be made in respect of any omission of work.

Valuation of additional work   **6.2**   In determining a fair and reasonable sum under Clause 6.1 for additional work the Engineer shall have regard to the prices contained in the Contract.

7 PAYMENT

Valuation of the Works   **7.1**   The Works shall be valued as provided for in the Contract.

Monthly statements   **7.2**   The Contractor shall submit to the Engineer at intervals of not less than one month a statement showing the estimated value of the Works executed up to the end of that period together with a list of any goods or materials delivered to the Site and their value and any other items which the Contractor considers should be included in an interim certificate.

Interim payments   **7.3**   Within 28 days of the delivery of such statement the Engineer shall certify and the Employer shall pay to the Contractor such sum as the Engineer considers is properly due less retention at the rate of and up to the limit set out in the Appendix. Until practical completion of the whole of the Works the Engineer shall not be required to certify any payment less than the sum stated in the Appendix as the minimum amount of interim certificate. The Engineer may by any certificate delete correct or modify any sum previously certified by him.

Payment of retention money   **7.4**   One half of the retention money shall be certified by the Engineer and paid to the Contractor within 14 days after the date on which the Engineer issues a Certificate of Practical Completion of the whole of the Works.

7.5   The remainder of the retention money shall be paid to the Contractor within 14 days after the issue of the Engineer's certificate under Clause 5.4.

Contractor to submit final account   **7.6**   Within 28 days after the issue of the Engineer's certificate under Clause 5.4 the Contractor shall submit a final account to the Engineer together with any documentation reasonably required to enable the Engineer to ascertain the final contract value. Within 42 days after the receipt of this information the Engineer shall issue the final certificate. The Employer shall pay to the Contractor the amount due thereon within 14 days of the issue of the final certificate.

7.7   The final certificate shall save in the case of fraud or dishonesty relating to or affecting any matter dealt with in the certificate be conclusive evidence as to the sum due to the Contractor under or arising out of the Contract (subject only to Clause 7.9) unless either party has within 28 days after the issue of the final certificate given notice under Clause 11.2.

Interest on overdue payments   **7.8**   In the event of failure by the Engineer to certify or the Employer to make payment in accordance with the Contract or any finding of an arbitrator to such effect the Employer shall pay to the Contractor interest compounded monthly on the amount which should have been certified or paid on a daily basis at a rate equivalent to 2% per annum above the base lending rate of the bank specified in the Appendix.

| **Value Added Tax** | **7.9** | (1) The Contractor shall be deemed not to have allowed in his tender for the tax payable by him as a taxable person to the Commissioners of Customs and Excise being tax chargeable on any taxable supplies to the Employer which are to be made under the Contract. |

(2) All certificates issued by the Engineer under Clauses 7.3 to 7.7 shall be net of Value Added Tax.

(3) In addition to the payments due under such certificates the Employer shall separately identify and pay to the Contractor any Value Added Tax properly chargeable by the Commissioners of Customs and Excise on the supply to the Employer of any goods and/or services by the Contractor under the Contract.

8 ASSIGNMENT AND SUB-CONTRACTING

Assignment **8.1** Neither the Employer nor the Contractor shall assign the Contract or any part thereof or any benefit or interest therein or thereunder without the written consent of the other party which consent shall not unreasonably be withheld.

No sub-contracting **8.2** The Contractor shall not sub-contract the whole of the Works. The Contractor
without shall not sub-contract any part of the Works without the consent of the
Engineer's Engineer which consent shall not unreasonably be withheld.
consent

Contractor **8.3** The Contractor shall be responsible for any acts defaults or neglects of any
responsible for sub-contractor his agents servants or workmen in the execution of the Works
sub-contractors or any part thereof as if they were the acts defaults or neglects of the
Contractor.

9 STATUTORY OBLIGATIONS

Contractor to **9.1** The Contractor shall subject to Clause 9.3 comply with and give all notices
comply with required by any statute statutory instrument rule or order or any regulation or
statutory by-law applicable to the construction of the Works (hereinafter called "the
requirements statutory requirements") and shall pay all fees and charges which are payable
in respect thereof.

Employer to **9.2** The Employer shall be responsible for obtaining in due time any consent
obtain consents approval licence or permission but only to the extent that the same may be
necessary for the Works in their permanent form.

Contractor's **9.3** The Contractor shall not be liable for any failure to comply with the statutory
exemption from requirements where and to the extent that such failure results from the
liability to Contractor having carried out the Works in accordance with the Contract or
comply with with any instruction of the Engineer.
statutes

10 LIABILITIES AND INSURANCE

Insurance of the **10.1** (1) If so stated in the Appendix the Contractor shall maintain insurance in the
Works joint names of the Employer and the Contractor in respect of the Works
(including for the purpose of Clause 10 any unfixed materials or other things
delivered to the Site for incorporation therein) to their full value against all
loss or damage from whatever cause arising (other than the Excepted Risks)
for which he is responsible under the terms of the Contract.

(2) Such insurance shall be effected in such a manner that the Employer and Contractor are covered for the period stipulated in Clause 3.2 and are also covered for loss or damage arising during the Defects Correction Period from such cause occurring prior to the commencement of the Defects Correction Period and for any loss or damage occasioned by the Contractor in the course of any operation carried out by him for the purpose of complying with his obligations under Clauses 4.7 and 5.2.

(3) The Contractor shall not be liable to insure against the necessity for the repair or reconstruction of any work constructed with materials or workmanship not in accordance with the requirements of the Contract.

(4) Any amounts not insured or not recovered from insurers whether as excesses carried under the policy or otherwise shall be borne by the Contractor or the Employer in accordance with their respective responsibilities under Clauses 3.2 and 3.3.

Contractor to indemnify the Employer **10.2** The Contractor shall indemnify and keep the Employer indemnified against all losses and claims for injury or damage to any person or property whatsoever (save for the matters for which the Contractor is responsible under Clause 3.2) which may arise out of or in consequence of the Works and against all claims demands proceedings damages costs charges and expenses whatsoever in respect thereof or in relation thereto subject to Clauses 10.3 and 10.4.

10.3 The liability of the Contractor to indemnify the Employer under Clause 10.2 shall be reduced proportionately to the extent that the act or neglect of the Engineer or the Employer his servants his agents or other contractors not employed by the Contractor may have contributed to the said loss injury or damage.

10.4 The Contractor shall not be liable for or in respect of or to indemnify the Employer against any compensation or damage for or with respect to

(a) damage to crops being on the Site (save in so far as possession has not been given to the Contractor)

(b) the use or occupation of land (which has been provided by the Employer) by the Works or any part thereof or for the purpose of constructing completing and maintaining the Works (including consequent loss of crops) or interference whether temporary or permanent with any right of way light air or water or other easement or quasi easement which are the unavoidable result of the construction of the Works in accordance with the Contract

(c) the right of the Employer to construct the Works or any part thereof on over under in or through any land

(d) damage which is the unavoidable result of the construction of the Works in accordance with the Contract

(e) death or injury to persons or loss of or damage to property resulting from any act or neglect or breach of statutory duty done or committed by the Engineer or the Employer his agents servants or other contractors (not being employed by the Contractor) or for or in respect of any claims demands proceedings damages costs charges and expenses in respect thereof or in relation thereto.

Employer to indemnify Contractor **10.5** The Employer shall save harmless and indemnify the Contractor from and against all claims demands proceedings damages costs charges and expenses in respect of the matters referred to in Clause 10.4. Provided always that the

11

Employer's liability to indemnify the Contractor under paragraph (e) of Clause 10.4 shall be reduced proportionately to the extent that the act or neglect of the Contractor or his sub-contractors servants or agents may have contributed to the said injury or damage.

Employer to approve insurance 10.6 The Contractor shall throughout the execution of the Works maintain insurance against damage loss or injury for which he is liable under Clause 10.2 subject to the exceptions provided by Clauses 10.3 and 10.4. Such insurance shall be effected with an insurer and in terms approved by the Employer (which approval shall not be unreasonably withheld) for at least the amount stated in the Appendix. The terms of such insurance shall include a provision whereby in the event of any claim in respect of which the Contractor would be entitled to receive indemnity under the policy being brought or made against the Employer the insurer will indemnify the Employer against any such claims and any costs charges and expenses in respect thereof.

Contractor to produce policies of insurance 10.7 Both the Employer and the Contractor shall comply with the terms of any policy issued in connection with the Contract and shall whenever required produce to the Employer the policy or policies of insurance and the receipts for the payment of the current premiums.

11 DISPUTES

Settlement of disputes 11.1 If any dispute or difference of any kind whatsoever shall arise between the Employer and the Contractor in connection with or arising out of the Contract or the carrying out of the Works (excluding a dispute under Clause 7.9 but including a dispute as to any act or omission of the Engineer) whether arising during the progress of the Works or after their completion it shall be settled according to the following provisions.

Notice of Dispute 11.2 For the purpose of Clauses 11.3 to 11.5 a dispute is deemed to arise when one party serves on the other a notice in writing (herein called the Notice of Dispute) stating the nature of the dispute. Provided that no Notice of Dispute may be served unless the party wishing to do so has first taken any step or invoked any procedure available elsewhere in the Contract in connection with the subject matter of such dispute and the other party or the Engineer as the case may be has

  (a) taken such steps as may be required

or

  (b) been allowed a reasonable time to take any such action.

Conciliation 11.3 In relation to any dispute notified under Clause 11.2 and in respect of which no Notice to Refer under Clause 11.5 has been served either party may within 28 days of the service of the Notice of Dispute give notice in writing requiring the dispute to be considered under the Institution of Civil Engineers' Conciliation Procedure (1994) or any amendment or modification thereof being in force at the date of such notice and the dispute shall thereafter be referred and considered in accordance with the said procedure.

Arbitration 11.4 Where a dispute has been referred to a conciliator under the provisions of Clause 11.3 either party may within 28 days of the receipt of the conciliator's recommendation refer the dispute to the arbitration of a person to be agreed upon by the parties by serving on the other party a written Notice to Refer. Where a written Notice to Refer is not served within the said period of 28 days the recommendation of the conciliator shall be deemed to have been accepted in settlement of the dispute.

11.5 Where a dispute has not been referred to a conciliator under the provisions of Clause 11.3 then either party may within 28 days of the service of the Notice of Dispute under Clause 11.2 refer the dispute to the arbitration of a person to be agreed upon by the parties by serving on the other party a written Notice to Refer. Where a Notice to Refer is not served within the said period of 28 days the Notice of Dispute shall be deemed to have been withdrawn.

**Appointment of 11.6
arbitrator** If the parties fail to appoint an arbitrator within 28 days of either party serving on the other party a written Notice to Concur in the appointment of an arbitrator the dispute shall be referred to a person to be appointed on the application of either party by the President (or if he is unable to act by any Vice President) for the time being of the Institution of Civil Engineers.

**ICE Arbitration 11.7
Procedure (1983)** Any such reference to arbitration shall be conducted in accordance with the Institution of Civil Engineers Arbitration Procedure (1983) or any amendment or modification thereof being in force at the time of the appointment of the arbitrator and unless otherwise agreed in writing shall follow the rules for the Short Procedure in Part F thereof provided that Rule 21.1 of the said Short Procedure shall not apply unless both parties agree thereto in writing after service of the Notice of Dispute provided also that an appointment under Clause 11.6 shall not be invalidated by failure to follow the timetable prescribed in Part A of the said Arbitration Procedure. Such arbitrator shall have full power to open up review and revise any decision instruction direction certificate or valuation of the Engineer.

12 APPLICATION TO SCOTLAND AND NORTHERN IRELAND

**Application to 12.1
Scotland** If the Works are situated in Scotland the Contract shall in all respects be construed and operate as a Scottish contract and shall be interpreted in accordance with the law of Scotland and Clause 12.2 shall apply.

12.2 In Clauses 11.1 to 11.7 hereof

(a) the word "arbiter" shall be substituted for the word "arbitrator"

(b) for any reference to the Institution of Civil Engineers Arbitration Procedure (1983) there shall be substituted a reference to the Institution of Civil Engineers Arbitration Procedure (Scotland)(1983)

and

(c) notwithstanding any of the other provisions of these Conditions nothing therein shall be construed as excluding or otherwise affecting the right of a party to arbitration to call in terms of Section 3 of the Administration of Justice (Scotland) Act 1972 for the arbiter to state a case.

**Application to 12.3
Northern Ireland** If the Works are situated in Northern Ireland the Contract shall in all respects be construed and operate as a Northern Irish contract and shall be interpreted in accordance with the law of Northern Ireland.

13 CONSTRUCTION (DESIGN AND MANAGMENT) REGULATIONS 1994

Definitions 13.1 In this Clause

(a) "the Regulations" means the Construction (Design and Management) Regulations 1994 or any statutory re-enactment or amendment thereof for the time being in force

(b) "Planning Supervisor" and "Principal Contractor" mean the persons so described in regulation 2(1) of the Regulations

(c) "Health and Safety Plan" means the plan prepared by virtue of regulation 15 of the Regulations.

Action to be taken **13.2** Where and to the extent that the Regulations apply to the Works and

(a) the Engineer is appointed Planning Supervisor

and/or

(b) the Contractor is appointed Principal Contractor

then in taking any action as such they shall state in writing that the action is being taken under the Regulations.

13.3 (1) Any action under the Regulations taken by either the Planning Supervisor or the Principal Contractor and in particular any alteration or amendment to the Health and Safety Plan shall be deemed to be an Engineer's instruction pursuant to Clause 2.3. Provided that the Contractor shall in no event be entitled to any additional payment and/or extension of time in respect of any such action to the extent that it results from any action lack of action or default on the part of the Contractor.

(2) If any such action of either the Planning Supervisor or the Principal Contractor could not in the Contractor's opinion reasonably have been foreseen by an experienced contractor the Contractor shall as early as practicable give written notice thereof to the Engineer.

14

ICE Conditions of Contract for Minor Works

APPENDIX TO THE CONDITIONS OF CONTRACT
(to be prepared before tenders are invited and to be included with the documents supplied to prospective tenderers)

1. Short description of the work to be carried out under the Contract

    ....

    ....

    ....

2. The payment to be made under Article 2 of the Agreement in accordance with Clause 7 will be ascertained on the following basis. (The alternatives not being used are to be deleted. Two or more bases for payment may be used on one Contract.)

    (a) Lump sum

    (b) Measure and value using a priced Bill of Quantities

    (c) Valuation based on a Schedule of Rates (with an indication in the Schedule of the approximate quantities of major items)

    (d) Valuation based on a Daywork Schedule

    (e) Cost plus fee (the cost is to be specifically defined in the Contract and will exclude off-site overheads and profit)

3. Where a Bill of Quantities or a Schedule of Rates is provided the method of measurement used is

    ....

4. Name of Engineer (Clause 2.1)

    ....

5. Starting date (if known) (Clause 4.1)

    ....

6. Period for completion (Clause 4.2)

    ....

7. Period for completion of parts of the Works if applicable and details of the work to be carried out within each such part (Clause 4.2)

	Details of work	Period for completion
Part A	....	....
Part B	....	....
Part C	....	....

8. Liquidated damages (Clause 4.6)

....

9. Limit of liquidated damages (Clause 4.6)

....

10. Defects Correction Period (Clause 5.1)

....

11. Rate of retention (Clause 7.3)

....

12. Limit of retention (Clause 7.3)

....

13. Minimum amount of interim certificate (Clause 7.3)

....

14. Bank whose base lending rate is to be used (Clause 7.8)

....

15. Insurance of the Works (Clause 10.1)
    Required/Not required

16. Minimum amount of third party insurance (persons and property) (Clause 10.6)

....
                              Any one accident/Number of accidents unlimited

17. Name of the Planning Supervisor (Clause 13(1)(b))

....

    Address ....

18. Name of the Principal Contractor (Clause 13(1)(b))

....

    Address ....

ICE Conditions of Contract
for Minor Works
Second Edition

GUIDANCE NOTES

1 The ICE Conditions of Contract for Minor Works are intended for use on contracts where

(a) the potential risks involved in the Works for both the Employer and the Contractor are adjudged to be small

(b) the Works are of a simple and straightforward nature

(c) the design of the Works, save for any design work for which the Contractor is made responsible (see Note 6), is complete in all essentials before tenders are invited

(d) the Contractor has no responsibility for the design of the Permanent Works other than possibly design of a specialist nature (see Note 6)

(e) nominated sub-contractors are not employed (but see Note 7)

(f) the Contract value does not exceed £250 000 and the period for completion of the Contract does not exceed 6 months except where the method of payment is on either a daywork or a cost plus fee basis.

2 The Contract Schedule should list all documents that will form part of the Contract. It is particularly important to ensure that the Appendix to the Conditions of Contract is prepared before tenders are invited (and the Appendix must then be included with the documents supplied to prospective tenderers). Notes on the completion of the Appendix are included in these guidance notes (see Note 13).

3 The method of payment for the Contract should be stated in the Appendix to the Conditions of Contract but if a Bill of Quantities is used the method of measurement used must also be indicated in the Appendix. If a Daywork Schedule other than the 'Schedules of Daywork carried out incidental to Contract Work' issued by the Federation of Civil Engineering Contractors is to be used the Schedule to be used must be clearly identified in the tender documents.

4 No provision is made for price fluctuations in view of the probable short duration of contracts let under these conditions. The letting of contracts on a daywork or cost plus fee basis is however not precluded (see Note 1(f)).

5 The procedures for letting and administering a minor works contract are intended to be as simple as possible in line with the low risk involved. The Contract should be fully defined in the documents listed in the Contract Schedule. There is no provision for amendment or addition to the Conditions of Contract and this should be avoided.

6 If the Contractor is required to be responsible for design work of a specialist nature which would normally be undertaken by a specialist sub-contractor or supplier (such as structural steelwork, mechanical equipment or an electrical or plumbing installation) full details must be given either in the Specification or in the Appendix to the Conditions of Contract or on the Drawings. The Contractor's responsibility in respect of such work shall be limited to the exercise of all reasonable skill care and diligence (Clause 7.3(c)).

7 The Engineer may in respect of any work that is to be sub-contracted or material purchased in connection with the Contract list in the Specification the names of approved sub-contractors or approved suppliers of material. Nothing however should prevent the Contractor from carrying out such work himself if he so chooses or from using other sub-contractors or suppliers of his own choice provided their workmanship or product is satisfactory and equal to that from an approved sub-contractor or supplier.

8 It is intended that acceptance should follow within 2 months of the date for submission of tenders.

9 Access as necessary to the Site should be available at the starting date under Clause 4.1.

10 In respect of Clause 2.2 it should be noted that in all normal circumstances the Engineer would not be expected to delegate his powers under Clauses 3.8, 4.4, 5.4 and 7.6.

11 In respect of Clause 3.4 it has to be recognized that in a minor works contract the Contractor might have no full-time supervisor on site and the Contractor may ask for instructions to be delivered or sent elsewhere for the attention of his representative. In these circumstances the Contractor has to accept the fact that urgent instructions might in the interests of safety or for some other reason have to be given directly to the Contractor's operatives on site.

12 In completing the Appendix to the Conditions of Contract the following points should be noted.

(1) Clause 2.1 (Name of the Engineer). It is the intention that the name of the Engineer who will be personally responsible for the Works should be stated.

(2) Clause 4.2 (Period for completion). If the Contract requires completion of parts of the Works by specified dates or within specified times such dates or times and details of the work involved in each part must be entered in the Appendix to the Conditions of Contract.

(3) Clause 4.6 (Liquidated damages and limit of liquidated damages). A genuine pre-estimate of the likely damage caused by any delay should be assessed and reduced to a daily or weekly rate. The limit of liquidated damages should not exceed 10% of the estimated final contract value and this should be taken into account when assessing the daily or weekly rate.

(4) Clause 5.1 (Defects Correction Period). This should normally be 6 months and in no case should exceed 12 months.

(5) Clause 7.3 (Rate of retention and limit of retention). The rate of retention should normally be 5%. A limit of retention has to be inserted in the Appendix and this should normally be between the limits of $2\frac{1}{2}$% and 5% of the estimated final Contract value.

(6) Clause 7.3 (Minimum interim certificate). It is recommended that the minimum amount of an interim certificate should be 10% of the estimated final Contract value rounded off upwards to the nearest £1000. This minimum only applies up to the date of practical completion of the whole of the Works.

(7) Clause 10.1 (Insurance of the Works). This is at the option of the Employer. It must be borne in mind that Contractors frequently carry large excesses on their all-risks policies so that the Contractor then accepts the risk under Clause 10.1 in respect of any uninsured loss. When the insurance under Clause 10.1 is to be provided by the Employer the details of such insurance, including any excesses which the Contractor may be expected to carry, should be stated in the tender documents.

(8) Clause 10.6 (Third party insurance). A minimum cover of £2 million for any one accident/unlimited number of accidents should normally be insisted upon. In certain locations where there is greater risk to adjacent properties a higher limit may be desirable.

13 The following points relating to disputes should be noted.

(1) The option provided by the Conciliation Procedure under Clause 11.3 is intended to provide a means whereby disputes can be settled with a minimum of delay by obtaining an independent recommendation as to how the matter in dispute should be settled. The conciliation is complete when the conciliator has delivered his recommendation and, if any, his opinion to the parties.

(2) It is normally expected that the party serving a Notice of Dispute under Clause 11.2 will at the same time serve notice in writing either under Clause 11.3 (Conciliation) or Clause 11.5 (Arbitration).

(3) It should be noted that if within the prescribed period of 28 days after serving a Notice of Dispute neither party has made a request in writing for the dispute to be referred to a conciliator nor has served a written Notice to Refer requiring the dispute to be referred to arbitration then the Notice of Dispute becomes void. It is then open to either party to continue the dispute by serving a fresh Notice of Dispute unless the first Notice of Dispute was served within the 28 days allowed under Clause 7.7 (Final certificate) in which case the final certificate becomes final and binding and no further dispute in respect of the Contract is possible.

Printed in Great Britain by Spottiswoode Ballantyne Printers Ltd, Colchester.

PRINCIPAL CHANGES FROM THE FIRST EDITION OF THE CONDITIONS OF CONTRACT FOR MINOR WORKS

Clauses 1.3, 1.4 and 1.5 have been revised in line with the wording in the ICE Conditions of Contract Sixth Edition and the ICE Design and Construct Conditions of Contract.

Clause 3.7 has been split into sub-clauses (a) and (b) and a new sub-clause (c) has been added which deals with design responsibility.

Clause 3.8 has been split into two sub-clauses and in (b) the wording has been revised to separate the requirements for the certification (by the Engineer) and the payment (by the Employer).

Clause 4.6. An additional paragraph has been added so that the Clause wording is now consistent with the wording in the ICE Conditions of Contract Sixth Edition (Clause 47(6)) and the ICE Design and Construct Conditions of Contract.

Clause 7.8. There has been an alteration in the wording so that the Clause will be consistent with the wording in the ICE Conditions of Contract Sixth Edition (Clause 60(7)) and the ICE Design and Construct Conditions of Contract.

Clause 7.9. The original Clause has been deleted and replaced by Clauses 70(1) and 70(2) from the ICE Conditions of Contract Sixth Edition and the ICE Design and Construct Conditions of Contract which are renumbered as Clauses 7.9(1) and 7.9(2).

Clause 10.1(4). A new sub-clause has been added so that Clause 10.1 is now consistent with the wording in the ICE Conditions of Contract Sixth Edition (Clause 21(2)(d)) and the ICE Design and Construct Conditions of Contract.

Clause 11.7 has been revised by the insertion of two sentences.

Clause 12.2 has been altered so that it is now worded in a similar way to the ICE Conditions of Contract Sixth Edition (Clause 60(7)) and the ICE Design and Construct Conditions of Contract.

Clause 12.3 has been added to cover application to Northern Ireland.

Clause 13 has been added to cover the requirements of the Construction (Design and Management) Regulations 1994.

Guidance Note 1. The order of the sub-clauses has been changed and the recommended limit on contract value has been increased to £250 000.

Guidance Notes 12 and 13 are the original Guidance Notes 13 and 14.

The original Guidance Note 12 has been omitted because this Form of Contract makes no provision for vesting.

Guidance Note 12(8). The recommended minimum cover for third party insurance has been increased to £2 million.

The Institution of Civil Engineers

CONCILIATION PROCEDURE (1994)

The Institution of Civil Engineers, The Association of Consulting Engineers and The Federation of Civil Engineering Contractors have, as sponsoring authorities, approved this document, commonly to be known as the ICE Conciliation Procedure.

A permanent joint committee of the sponsoring authorities will keep under review the use of this document and will consider any suggestion for amendment, which should be addressed to the Secretary, The Institution of Civil Engineers, 1–7 Great George Street, London SW1P 3AA. Revision to the document will be made when such action seems warranted.

From: ICE Conditions of Contract for Minor Works 2nd Ed. Acc. No. 11061529

Published for The Institution of Civil Engineers by Thomas Telford Services Ltd, Thomas Telford House, 1 Heron Quay, London E14 4JD

Copies may also be obtained from The Secretary, The Association of Consulting Engineers, Alliance House, 12 Caxton Street, London SW1H 0QL, from The General Secretary, The Federation of Civil Engineering Contractors, Cowdray House, 6 Portugal Street, London WC2 2HH, or, for those calling in person, from The Institution of Civil Engineers, 1–7 Great George Street, London SW1P 3AA.

The Institution of Civil Engineers

CONCILIATION PROCEDURE

PREFACE

The main difference between Conciliation and Arbitration is that the outcome of a conciliation is not imposed and only becomes binding with the consent of each party. Conciliation therefore allows the parties to the dispute the freedom to explore ways of settling the dispute with the assistance of an independent impartial person—the Conciliator.

Conciliation is essentially an assisted negotiation.

The aim of Conciliation is to reach an agreed solution. If for any reason that does not prove possible then the Conciliator is required to make a Recommendation as to how, in his opinion, the matter should be settled.

Because the aim is to reach a settlement, the Conciliation proceedings are conducted on a 'without prejudice' basis. This means that nothing disclosed during the Conciliation may be used as evidence in any subsequent proceedings, whether in arbitration or litigation.

The Procedure permits the Conciliator to communicate privately and separately with each party without subsequently revealing to any other party what he has been told; this is something not permitted to an arbitrator or judge.

The Conciliator's job is to explore with the parties their interests, strengths and weaknesses, and perceived needs; to identify possible areas of accommodation or compromise, and to search for possible alternative solutions. Anything can be explored which could lead the parties to an agreed settlement. Where the Conciliation follows an Engineer's decision, the parties are equally free to explore options that were not available to the Engineer.

The information given to the Conciliator is not comprehensive, is not given under oath or affirmation, and there is no cross-examination. The Conciliator is not bound by the rules of natural justice or the rules of evidence and can be guided only by what the parties choose to tell him and his own professional knowledge and experience.

If the parties can reach a commercial settlement the Conciliator will assist the parties to ensure that any such agreement is recorded in writing in a form that is enforceable. If requested, the Conciliator may be appointed by the parties as an arbitrator with authority solely to issue a 'Consent Award' which will make any settlement reached immediately enforceable.

If agreement cannot be reached the Conciliator will make a Recommendation.

Because the information given to the Conciliator is selective, it follows that the Recommendation is not comparable with the award of an arbitrator or the judgement of a court. The Recommendation is the Conciliator's opinion of how the dispute might be resolved in the most practical way.

CONCILIATION UNDER THE ICE CONTRACTS

Conciliation may be used at any time, by the agreement of the parties, to resolve disputes that arise, and it is a stage in the dispute resolution procedures in the following contracts

Contracts published by the Institution of Civil Engineers
ICE Conditions of Contract (Sixth Edition) [ICE 6]: optional
ICE Design and Construct Contract [D&C]: mandatory
ICE Minor Works Contract [MWC]: optional

Contracts published by the Federation of Civil Engineering Contractors
Form of Subcontract (September 1991) for use with the ICE Conditions of
   Contract (Sixth Edition): optional
Form of Subcontract (revised September 1984) as amended for use with the
   ICE Minor Works Contract: optional.

Conciliation may be used with other standard contracts with the agreement of the
parties. This agreement may be made either before or after the Contract is formed.

If it is agreed to refer a dispute to Conciliation under a contract which does not
include a conciliation clause it is important to check the timetable under the
dispute resolution clauses to ensure that any time limits stated therein will not be
exceeded because of the Conciliation. If there is such a likelihood the agreement
to conciliate should include provision for suspending such a timetable for the
period of the Conciliation.

The Institution of Civil Engineers

CONCILIATION PROCEDURE (1994)

1   This Procedure shall apply whenever

    (a) the Parties have entered into a contract which provides for Conciliation for any dispute which may arise between the Parties in accordance with the Institution of Civil Engineers' Conciliation Procedure, or

    (b) where the Parties have agreed that the Institution of Civil Engineers' Conciliation Procedure shall apply.

2   This Procedure shall be interpreted and applied in the manner most conducive to the efficient conduct of the proceedings with the primary objective of achieving a settlement to the dispute by agreement between the Parties as quickly as possible.

3   Subject to the provisions of the Contract relating to Conciliation, any Party to the Contract may by giving to the other Party a written notice, hereafter called a Notice of Conciliation, request that any dispute in connection with or arising out of the Contract or the carrying out of the Works shall be referred to a Conciliator. Such Notice shall be accompanied by a brief statement of the matter or matters which it is desired to refer to Conciliation, and the relief or remedy sought.

4   Save where a Conciliator has already been appointed, the Parties shall agree upon a Conciliator within 14 days of the Notice being given under Paragraph 3. In default of agreement any Party may request the President (or, if he is unable to act, any Vice President) for the time being of the Institution of Civil Engineers to appoint a Conciliator within 14 days of receipt of the request by him, which request shall be accompanied by a copy of the Notice of Conciliation.

5   If, for any reason whatsoever, the Conciliator is unable, or fails to complete the Conciliation in accordance with this Procedure, then any Party may require the appointment of a replacement Conciliator in accordance with the procedures of Paragraph 4.

6   The Party requesting Conciliation shall deliver to the Conciliator, immediately on his appointment, and at the same time to the other Party if this has not already been done, a copy of the Notice of Conciliation, or as otherwise required by the Contract, together with copies of all relevant Notices of Dispute and of any other notice or decision which is a condition precedent to Conciliation.

7   The Conciliator shall start the Conciliation as soon as possible after his appointment and shall use his best endeavours to conclude the Conciliation as soon as possible and in any event within any time limit as may be stated in the Contract, or two months from the date of his appointment, or within such other time as may be agreed between the Parties.

8   Any Party may, upon receipt of notice of the appointment of the Conciliator and within such period as the Conciliator may allow, send to the Conciliator and to the other Party a statement of its views on the dispute and any issues that it considers to be of relevance to the dispute, and any financial consequences.

9 As soon as possible after his appointment, the Conciliator shall issue instructions establishing, *inter alia*, the date and place for the conciliation meeting with the Parties. Each Party shall inform the Conciliator in writing of the name of its representative for the Conciliation, who shall have full authority to act on behalf of that Party, and the names of any other persons who will attend the Conciliation meeting. This information shall be given at least seven days before the Conciliation meeting with copies to the other Party.

10 The Conciliator may, entirely at his own discretion, issue such further instructions as he considers to be appropriate, meet and question the Parties and their representatives, together or separately, investigate the facts and circumstances of the dispute, visit the site and request the production of documents or the attendance of people whom he considers could assist in any way. The Conciliator may conduct the proceedings in any way that he wishes, and with the prior agreement of the Parties obtain legal or technical advice, the cost of which shall be met by the Parties, in accordance with Paragraph 17, or as may be agreed by the Parties and the Conciliator.

11 The Conciliator may consider and discuss such solutions to the dispute as he thinks appropriate or as may be suggested by any Party. He shall observe and maintain the confidentiality of particular information which he is given by any Party privately, and may disclose it only with the explicit permission of that Party. He will try to assist the Parties to resolve the dispute in any way which is acceptable to them.

12 Any Party may, at any time, ask that additional claims or disputes, or additional parties, shall be joined in the Conciliation. Such applications shall be accompanied by details of the relevant contractual facts, notices and decisions. Such joinder shall be subject to the agreement of the Conciliator and all other Parties. Any additional party shall, unless otherwise agreed by the Parties, have the same rights and obligations as the other Parties to the Conciliation.

13 If, in the opinion of the Conciliator, the resolution of the dispute would be assisted by further investigation by any Party or by the Conciliator, or by an interim agreement, including some action by any Party, then the Conciliator will, with the agreement of the Parties, give instructions and adjourn the proceedings as may be appropriate.

14 Once a settlement has been achieved of the whole or any part of the matters in dispute, the Conciliator may assist the Parties to prepare an Agreement incorporating the terms of the settlement.

15 If, in the opinion of the Conciliator, it is unlikely that the Parties will achieve an agreed settlement to their disputes, or if any Party fails to respond to an instruction by the Conciliator, or upon the request of any Party, the Conciliator may advise all Parties accordingly and will forthwith prepare his Recommendation.

16 The Conciliator's Recommendation shall state his solution to the dispute which has been referred for Conciliation. The Recommendation shall not disclose any information which any Party has provided in confidence. It shall be based on his opinion as to how the Parties can best dispose of the dispute between them and need not necessarily be based on any principles of the Contract, law, or equity. The Conciliator shall not be required to give reasons for his Recommendation. Nevertheless should he choose to do so, his reasons shall be issued as a separate document, within 7 days of the giving of his Recommendation.

17 When a settlement has been reached or when the Conciliator has prepared his Recommendation, or at an earlier date solely at the discretion of the Conciliator, he shall notify the Parties in writing and send them an account of his fees and disbursements. Unless otherwise agreed between themselves each Party shall be responsible for paying and shall within 7 days of receipt of the account from the Conciliator pay an equal share save that the Parties shall be jointly and severally liable to the Conciliator for the whole of his account. Upon receipt of payment in full the Conciliator shall send his Recommendation to all the Parties. If any Party fails to make the payment due from him the other Party may pay the sum to the Conciliator and recover the amount from the defaulting Party as a debt due. Each Party shall meet his own costs and expenses.

18 The Conciliator may be recalled, by written agreement of the Parties and upon payment of an additional fee, to clarify, amplify or give further consideration to any provision of the Recommendation.

19 The Conciliator shall not be appointed arbitrator in any subsequent arbitration between the Parties whether arising out of the dispute, difference or other matter or otherwise arising out of the same Contract unless the Parties otherwise agree in writing. No Party shall be entitled to call the Conciliator as a witness in any subsequent arbitration or litigation concerning the subject matter of the Conciliation.

20 The confidential nature of the Conciliation shall be respected by every person who is involved in whatever capacity.

21 The Conciliator shall not be liable to the Parties or any person claiming through them for any matter arising out of or in connection with the Conciliation or the way in which it is or has been conducted, and the Parties will not themselves bring any such claims against him.

22 Any notice required under this Procedure shall be sent to the Parties by recorded delivery to the principal place of business or if a company to its registered office, or to the address which the Party has notified to the Conciliator. Any notice required by this Procedure to be sent to the Conciliator shall be sent by recorded delivery to him at the address which he shall notify to the Parties on his appointment.

23 In this Procedure where the context so requires 'Party' shall mean 'Parties' and 'he' shall mean 'she'.

The Institution of Civil Engineers

CONCILIATOR'S AGREEMENT

(for use when an appointment is made by the Parties)

THIS AGREEMENT is made on .... day of .... 19......

between .... (First Party)

of ....

and .... (Second Party)

of ....

and (where there is a third party) ....

of ....

(hereinafter called 'the Parties') of the one part and ....

of ....

(hereinafter called 'the Conciliator') of the other part.

Disputes or differences have arisen between the Parties in connection with certain construction works known as

....

and the Parties have agreed that these disputes or differences shall be considered under the Institution of Civil Engineers' Conciliation Procedure (1994) (hereinafter called 'the Procedure') and have agreed to ask the Conciliator to act.

IT IS NOW AGREED as follows

1   The Conciliator hereby accepts the appointment and agrees to conduct the Conciliation in accordance with the Procedure. He also agrees that, if requested or required to do so, he will prepare a Recommendation in accordance with Paragraphs 15 and 16 of the Procedure.

2   The Parties bind themselves jointly and severally to pay upon demand the Conciliator's fees and disbursements in accordance with Paragraph 17 of the Procedure in the manner set out in the attached Schedule.

3   The Parties and the Conciliator shall at all times maintain the confidentiality of this Conciliation and shall ensure that anyone acting on their behalf or through them will do likewise.

Signed on behalf of

First Party ....

in the presence of
(name of witness) ....

(address of witness) ....

....

Second Party ....

in the presence of
(name of witness) ....

(address of witness) ....

....

Third Party ....

in the presence of
(name of witness) ....

(address of witness) ....

....

Conciliator ....

in the presence of
(name of witness) ....

(address of witness) ....

....

The Institution of Civil Engineers

SCHEDULE TO THE CONCILIATOR'S AGREEMENT

1   The Conciliator shall be paid at the hourly rate of £ .... in respect of all time spent upon, or in connection with, the Conciliation including time spent travelling.

2   The Conciliator shall be reimbursed in respect of all disbursements properly made including, but not restricted to

   (a) printing, reproduction and purchase of all documents, drawings, maps, records and photographs
   (b) telegrams, telex, faxes, and telephone calls
   (c) postage and similar delivery charges
   (d) travelling, hotel expenses and other similar disbursements
   (e) room charges.

3   The Conciliator shall be paid an appointment fee of £ .... to be paid in equal amounts by each Party within 14 days of the appointment of the Conciliator. This fee will be deducted from the final statement of any sums which may become payable under Item 1 and/or Item 2 of this Schedule.

4   The Conciliator is/is not* registered for VAT.

5   When the Conciliator is registered for VAT, VAT shall be charged additionally in accordance with the rates current at the date of invoice.

*Delete whichever is not applicable.

ISBN 0 7277 20147